Queues
Will This Wait Never End!

Queues

Will This Wait Never End!

Clifford Sloyer
Wayne Copes
William Sacco
Robert Stark

JANSON PUBLICATIONS, INC. Providence, Rhode Island

Material based on work supported by the National Science Foundation and produced by the Committee on Enrichment Modules, Department of Mathematical Sciences, University of Delaware, and Tri-Analytics, Bel Air, Maryland.

94 93 92 91 90 89 88 87   8 7 6 5 4 3 2

Contents

The authors wish to thank Jane Melville, who helped to produce this monograph.

Preface

Everyone in the family is looking forward to a weekend at the seashore. The beach, the water, the boardwalk will all be great. Even the ride there will be fun. We're getting close to the Bay Bridge now. UH-OH! We're slowing down. Wonder why? All I can see is miles of tail lights inching toward the toll plaza. Too many cars! Not enough booths! Exact change lanes still almost empty. So slow! Getting restless! At last we're through the booth, and making good time again. We'll be there soon now. Wonder why they didn't plan that toll plaza better? Reminds me of the line in the cafeteria at lunchtime. Reminds me of the Friday night movie lines. Reminds me of . . . .

Part I
BASIC IDEAS

A *queue* is a waiting line. People form queues at checkout counters in super-markets, at telephone booths, and at ticket booths in theatres. A "customer" in a queue need not be a human being. For example, in an automobile production line, the cars waiting for final inspection form a queue with the waiting cars as "customers."

The following table lists several common situations together with queuing ingredients.

Situation	Customers	Service (Processing) Facility	Service Performed
Doctor's office	Patients	Doctor, staff, equipment	Medical treatment or examination
Auto traffic at a toll booth	Automobiles (Drivers)	Toll booth and collector or exact change machine	Collection of fees—passage through a certain point on a highway
Buying groceries at a supermarket	People with loaded grocery carts	Checkout counter	Obtaining total cost, accepting payment, and bagging groceries
Planes arriving at an airport	Planes	Runway and terminal	Landing, unloading, and loading
Making a telephone call from a booth	People wishing to telephone another person(s)	Phone booth	Communication with another person(s)
Buying tickets to a show at a theatre	Theatre goers	Ticket booth	Providing acceptable token for entrance into a theatre
Final inspection of newly produced automobiles	Automobiles	Inspection staff and equipment	Final judgment on quality of new car

1. List four other situations involving queues.

Situation	Customers	Service (Processing) Facility	Service Performed
a.			
b.			
c.			
d.			

One observes that there are many situations in daily life when a queue is formed if the service required by a customer is not immediately available. In many situations customers arrive on an irregular basis to demand service. For example, at a checkout counter in a local supermarket, more customers arrive for service late in the afternoon than arrive at earlier hours. The amount of time required to satisfy (service) customers also varies. For example, at a gas station, one customer may simply ask for a fill-up while the next customer asks for a fill-up, an oil check, and a windshield cleaning. The time required for a fill-up alone varies with the size of the car and the amount of gas previously in the tank. Thus queues may grow for a while, then disappear completely during a lull in demand (in which case we say that the queue length is zero), and then reappear still later.

Queues present problems of various types. One that should not be overlooked is simply the frustration of people in a queue. Another important problem is that of the cost involved with queues. For example, if a machine on an assembly line breaks down and a queue forms until repairs are made, the cost of lost production and idle time can be considerable. Consider the following three examples which involve problems presented by queues.

Example 1: A bank offers a large variety of services to the public: check cashing, bond cashing, savings account payments, savings account withdrawals, etc. Demand for these services varies through the day and during the week. The area in which people can wait for service is limited, so some customers may go away without being able to make the transaction they desired. This could result in customers seeking service with other banks. The time taken to complete a transaction varies. For example, cashing a bond normally requires more time than cashing a personal check. Is it better for the customers to have each kind of service available at different positions (tellers) or to have all positions providing all services? How many tellers should be available on Fridays? On Tuesdays? How long is an "average" person willing to wait for service?

Example 2: A communications center has a certain number of lines for telephone calls. One cannot predict when a customer will make a call. The customer, on the other hand, is not aware of other customers who may want to make a call (use a line) at the same time. If the lines are all being used when a customer(s) wishes to make a call and the customers(s) must wait until one is free, then a queue will form. The time of the conversations will vary from one customer to another. How many lines should this communications center have available in order to provide "acceptable" service to its customers? How does one determine a meaning of "acceptable" in this context?

Example 3: A certain post office has the capability of providing five clerks to work simultaneously. There is not enough work space for any additional clerks. Customers enter a common queue and are serviced by the first available clerk. The "Christmas Rush" normally occurs between November 20 and December 23. How many clerks should the postmaster have available during this period? What should be taken into account—cost or public opinion of the postal service? What is gained by having four clerks rather than three? Three clerks rather than two? etc. What does the word "gained" mean in this context?

Read ahead and learn how to solve many of the problems presented by queues.

Part II
CONSTANT ARRIVAL RATES AND SERVICE TIMES

There are many levels or degrees of complexity that can exist within queues. We will start our study with the simplest situation and use it to identify some basic concepts which are generally useful. This first situation is one in which:

(1) customers arrive at a constant rate,
(2) there is a single server,
(3) the server processes customers at a constant rate.

While this does represent a simple situation, there are many real-world examples of automated production methods which it does accurately represent.

One of the authors recently toured a metal-craft factory and observed a machine which was used to "smooth" the raw items (metal ashtrays and beer mugs, etc.). These items were then taken by a belt to another machine where a lacquer finish was sprayed on each item. One could regard the smooth items as customers for the spray machine and the spray machine as a service facility. In this case customers arrived at a constant rate with approximately 12 seconds between arrivals. The time required to spray (service) an item was approximately 7 seconds.

2. For these arrival and processing rates (of 1 arrival every 12 seconds and 1 item processed every 7 seconds) and assuming no breakdowns, what percent of the time is the sprayer actually working? _____

3. How many items are in the queue to the sprayer after 5 minutes of operation, if the item which arrived exactly 5 minutes after the operation started is ignored? _____

4. What is the fastest arrival rate of items to the spraying machine which will not allow any queue to build up? _____

5. At the fastest arrival rate, what percent of the time is the sprayer actually working? _____

6. Suppose one item arrives for processing every x seconds and processing each item requires y seconds. What happens to the queue if the following order relations hold between x and y?

   a. $x < y$ _____
   b. $x = y$ _____
   c. $x > y$ _____

7. If $x < y$ for a certain processor in a production system, name 2 ways to overcome the problem:

a. _____

b. _____

One way which may have come to mind is the addition of a second server or sprayer. In fact, it is through the analysis of queue formation and build-up that the appropriate number of check-out stations in supermarkets, post offices, booths in toll plazas, etc. are determined. More will be presented on those problems later.

8. Give two other examples where arrival times and service times are constant.

    a. _____

    b. _____

9. Suppose the spraying machine is given at least one hour of preventive maintenance (that is: clean the nozzle, spray the lacquer, etc.) each work day at 8 o'clock. The smoothing machine, however, not needing daily maintenance, starts its work at 8 o'clock. Surely a queue will form while the sprayer is being readied. (But experience shows it always catches up.) At two minutes past eight, there are 10 items in the queue. How many items are in this queue at the end of :

    a. 3 minutes? _____

    b. 10 minutes? _____

    c. 24 minutes? _____

Note that as a result of the breakdown, items must wait to be processed, just as you do in a supermarket or doctor's office. The customers in this example (items waiting to be painted) don't get impatient, but people customers do. Usually customers want to be served without waiting "too long," and if that time is exceeded they will take their business elsewhere. Thus, one important measure of service performance is the average time which customers must wait in line for service. More will be said about this measure shortly.

Items move from machine A to machine B at the rate of 30 items per hour. Machine B is capable of servicing 80 items per hour. If machine B breaks down, a queue is going to form (items waiting to get into the production process).

10. Suppose machine B is inoperative for 4 hours before it begins to operate again. Assuming that machine A continues to operate during this period, when machine B begins to operate again there will be a queue of 120 items. During the next hour, 30 new items will join the waiting line, but machine B will handle 80 items so that at the end of 1 hour there will be 70 items in the queue.

Complete the following table:

Time Factor	Waiting Line
Machine B begins	120
End of one hour	70
End of two hours	_____
End of three hours	_____

11. Suppose machine B requires 7 hours for repairs before operations can begin again. Assuming that machine A continues to operate during this period, after how many hours (an integral number) will there be no queue? _____

Hint: It might be helpful to complete the following table:

Time Factor	Waiting Line
Machine B begins	210
End of one hour	_____
End of two hours	_____
End of three hours	_____
(continue as necessary)	

12. Suppose machine B requires t hours for repairs before operations can begin again (t an integer). Assuming that machine A continues to operate during this period, when machine B begins to operate again there will be a queue of $30t$ items. During the next hour 30 new items will join the waiting line, but machine B will handle 80 items so that at the end of one hour there will be:

$$30t + 30 - 80$$

or

$$30(t + 1) - 80$$

or

$$30t - 50$$

items in the queue.
Complete the following table:

Time Factor	Waiting Line		
Machine B begins	$30t$	or	$30t$
End of one hour	$30(t + 1) - 1 \cdot (80)$	or	$30t - 1 \cdot (50)$
End of two hours	$30(t + 2) - 2 \cdot (80)$	or	$30t - 2(50)$
End of three hours	_____	or	_____
End of four hours	_____	or	_____
.	.		.
.	.		.
.	.		.
End of M hours	_____	or	_____

13. The smallest value of M (an integer) for which there is no queue is the smallest integer M for which $M \geq$ _____

A generalization: Items move from machine A to machine B at the rate of x items per hour. Machine B is capable of servicing y items per hour.

14. Why must one assume that $x \leq y$? _____
_____
_____

15. Machine B breaks down and t hours are required for repairs before operations can begin again. Assuming that machine A continues to operate during this period, when machine B begins to operate again there will be a queue of xt items. During the next hour x new items will join the waiting line, but machine B will handle y items so that at the end of one hour there will be

$$xt + x - y$$

or

$$x(t + 1) - y$$

or

$$xt - (y - x)$$

items in the queue.
Complete the following table:

Time Factor	Waiting Line		
Machine B begins	xt	or	xt
End of one hour	$x(t + 1) - y$	or	$xt - (y - x)$
End of two hours	_____	or	_____
End of three hours	_____	or	_____
End of four hours	_____	or	_____
.	.	.	.
.	.	.	.
.	.	.	.
End of M hours	_____	or	_____

16. The smallest value of M for which there is no queue is the smallest integer M for which $M \geq$ _____

Have you used the fact that $x < y$ in obtaining your result? What happens if $x = y$? _____

A. Graphical Descriptions of Service Facility

It will be useful to have a visible way to keep track of the customers arriving and being processed by a server. Let us start by drawing a time line, or axis, on which we will note the arrival and departure of customers.

Time (Minutes)

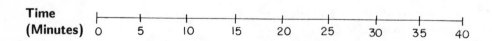

Time 0 may be taken to represent the "opening" of the serving facility. Suppose the first customer, denoted c_1, arrives at $t = 3$. We will indicate this event on the time line as follows:

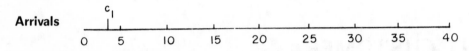

Arrivals

In fact, all arrivals will be noted, at the appropriate times, by labeled short vertical lines above the axis. If succeeding customers arrive at exactly 5 minute intervals, indicate their arrivals on the time line above.

Suppose that each customer is serviced in exactly 2 minutes. We denote the completed serving of a customer by a labeled vertical line extending below the time axis at the appropriate time. For example, since c_1 arrived at the server at $t = 3$ when there was no queue, he was immediately processed; his processing was completed at $t = 5$ and is shown as follows:

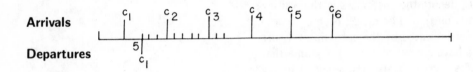

Arrivals

Departures

Indicate on the time line the departures of the other five arriving customers shown above, assuming that each customer is serviced in 2 minutes.

By now you should have a general understanding of the simplest sort of queue, one with constant arrival rates and service times, and how to graphically represent such a queue. We now go on to look at a specific problem in more detail.

Part III
THE FIRST NO-WAIT CUSTOMER

Consider a service facility which requires 2 minutes to service a customer and a new customer arrives every 5 minutes. When the service facility begins operation there are 6 customers waiting, and the first "new" customer arrives one minute later. Let e_1, e_2, e_3, e_4, e_5, e_6 denote the customers in the waiting line when the operation begins ("e" for early bird), and let c_1, c_2, c_3, etc. denote the new customers that arrive, in the order indicated by the subscripts. (Note that the early-birds may be considered to have arrived at $t = 0$.) Assume that customers are serviced on a first come–first served basis. (However, in many applications, e.g., emergencies arriving at a doctor's office, this assumption would not be a good one!) We will expand our graphical description of this situation to two time lines as shown below. In this way, the start of *service times* can be shown explicitly for each customer.

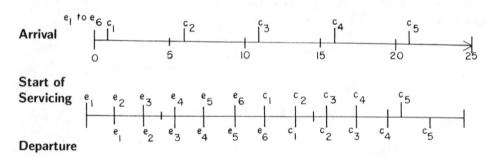

One observes that c_5 is the first customer to arrive with no queue present. Here the queue consists of the customers in the waiting line *or* in the service facility.

17. The number of minutes (time measured from the beginning of operation) required before no queue is present is _____

18. The total number of customers served before the first "no-wait" customer arrived is _____

19. When the first "no-wait" customer arrives, the time since the facility began operating is _____ minutes.

20. When the first "no-wait" customer arrives, the facility has already spent _____ minutes actually servicing customers.

After the server has overcome the backlog created by early-birds (or a breakdown), there is no longer any queue formation since service time is less than the time between arrivals. But as mentioned earlier, in certain applications a useful measure of system performance is average waiting time, which is computed by dividing the sum of the waiting times for the individual customers by the total number of customers. So let's examine the average waiting time for the customers processed before c_5 in our previous example.

21. Fill in the waiting times (time from arrival till start of processing) below.

Customer	e_1	e_2	e_3	e_4	e_5	e_6	c_1	c_2	c_3	c_4
Waiting Time	0	2	4	—	—	—	—	—	—	—

22. The average waiting time for the customers in the table above is _____

23. What is the average waiting time for customers c_5 through c_{10} _____

How long are you willing to wait in a supermarket line before becoming annoyed? _____

Wait to see a doctor? _____

Wait for a sales person at a department store? _____

A welding machine in a production line requires 4 minutes to service a customer. When operation begins there are 5 customers waiting in line. Two minutes after the beginning of operation, a new customer arrives and after that a new one arrives every 7 minutes. Use an appropriate time scale and the lines below to indicate arrival, start of service, and departure times.

Arrival

Start of Servicing

Departure

24. The first new customer to arrive and find no queue is _____

25. The total number of customers served before the first "no-wait" customer arrives is _____

26. What is the total amount of time the welding machine spent servicing customers before the first "no-wait" customer arrived (actual service time)?

_____

27. How many minutes after the welding machine began operating (available operating time) did the first "no-wait" customer arrive? _____

28. What is the average waiting time for the customers serviced before the first customer who arrives and faces no queue? _____

A. Actual Service Time

Suppose that a facility requires 3 minutes to service a customer. When the facility opens there are 4 customers waiting. The first customer to arrive and find no queue is the tenth new customer. When this customer arrives, the *actual service time* of the facility would be 39 minutes since 13 customers were previously serviced, each requiring three minutes.

29. A facility requires 4 minutes to service a customer. When the facility opens there are 5 customers waiting. The first customer to arrive and find no queue is the seventh new customer. When this customer arrives, the actual service time of the facility would be _____ minutes.

30. A facility requires 2 minutes to service a customer. When the facility opens there are 10 customers waiting. The first customer to arrive and find no queue is the eighth new customer. When this customer arrives, the actual service time of the facility would be _____ minutes.

31. A facility requires S minutes to service a customer. When the facility opens there are 14 customers waiting. The first customer to arrive and find no queue is the eleventh new customer. When this customer arrives, the actual service time of the facility would be _____ minutes.

32. A facility requires S minutes to service a customer. When the facility opens there are W people waiting. The first customer to arrive and find no queue is the seventh new customer. When this customer arrives, the actual service time of the facility would be _____ minutes.

33. A facility requires S minutes to service a customer. When the facility opens there are W people waiting. The first customer to arrive and find no queue is the $(n+1)$st new customer. When this customer arrives, the actual service time of the facility would be _____ minutes.

B. Available Operating Time

Suppose that new customers arrive at a facility every 4 minutes. The first new customer arrives 1 minute after the facility opens. The first customer to arrive and find no queue is the eighth new customer. When this customer arrives, the available operating time of the facility has been 29 minutes, since 7 four-minute intervals have elapsed since the arrival of the first new customer, one minute after opening.

34. New customers arrive at a facility every 5 minutes. The first new customer arrives 2 minutes after the facility opens. The first customer to arrive and find no queue is the seventh new customer. When this customer arrives, the available operating time of the facility has been _____ minutes.

35. New customers arrive at a facility every 10 minutes. The first new customer arrives 4 minutes after the facility opens. The first customer to arrive and find no queue is the fifth new customer. When this customer arrives, the available operating time of the facility has been _____ minutes.

36. New customers arrive at a facility every 7 minutes. The first new customer arrives F minutes after the facility opens. The first customer to arrive and find no queue is the ninth new customer. When this customer arrives, the available operating time has been _____ minutes.

37. New customers arrive at a facility every T minutes. The first new customer arrives F minutes after the facility opens. The first customer to arrive and find no queue is the tenth new customer. When this customer arrives, the available operating time has been _____ minutes.

38. New customers arrive at a facility every T minutes. The first new customer arrives F minutes after the facility opens. The first customer to arrive and find no queue is the $(n+1)$st new customer. When this customer arrives, the available operating time has been _____ minutes.

The reader should be sure that he or she understands that available operating time is *not* affected by the fact that customers are waiting in queues to be serviced.

39. What is the order relation (i.e., $<$, $>$, $\geq$, etc.) between actual service time and the available operating time? _____

C. **The General Case (or Let's Put It Together)**

We began this section with the following problem: A service facility requires 2 minutes to service a customer. A new customer arrives every 5 minutes. When the service facility begins operation there are 6 customers waiting and the first new customer arrives 1 minute later. Our problem was to determine which customer would be the first to arrive and find no queue.

A line indicating arrival and departure times was used to solve this problem. The use of such lines, however, can become cumbersome and impractical. Thus one seeks an analytic approach. Let c_{n+1} denote the first customer to arrive and find no queue present.

When this customer arrives, the actual service time of the facility has been

$$(n+6) \cdot 2$$

minutes.

Why? _____

_____

_____

The available operating time of the facility has been

$$n \cdot 5 + 1$$

minutes.

Why? _____

_____

_____

Hence we must have

$$(n+6) \cdot 2 \leq n \cdot 5 + 1$$
$$11 \leq 3n$$
$$\frac{11}{3} \leq n. \tag{1}$$

Note that n must be an integer. Moreover, we want the *first* customer to arrive and find no queue. Thus one wants the smallest integer n which satisfies (1). The result is $n = 4$ and it follows that the fifth new customer is the first to arrive and find no queue. This agrees with our former solution.

40. The actual service time before no queue was present was _____ minutes.

Let us generalize the above process by letting W denote the number of customers in the waiting line when the service facility opens, S the time required to service a customer, and T the time between new arrivals. Let F denote the time between the opening of the service facility and the arrival of the first new customer. We shall assume that $0 \leq F \leq WS$. In a later problem you will be asked why this assumption is made.

41. In order for the queue to disappear, the following order relation must hold between T and S: _____

Now let c_{n+1} denote the first customer to arrive and find no queue.

42. When customer c_{n+1} arrives, the actual service time of the facility has been _____ minutes.

43. When customer c_{n+1} arrives, the available operating time of the facility has been _____ minutes.

44. The results of questions 41 and 42 must satisfy the following order relation:

_____. (2)

45. Thus one must have

$$n \geq \text{_____}, \tag{3}$$

where the right-hand side of the inequality in (3) does not contain n.

46. The problem of finding the first "no-wait" customer has been solved. One simply

_____

_____

_____

47. The algebraic expression for actual service time required before no queue
was present was _____ minutes.

48. In going from the inequality in (2) to the inequality in (3) did you make use
of the order relation between S and T?
Where?

_____

_____

_____

49. In our generalization we assumed that $F < WS$ (as was the case in our
example). What happens if $F \geq WS$?

_____

_____

_____

Part IV
VARIABLE ARRIVALS AND SERVICE TIMES—SIMULATION

A. Random Number Tables

In the last section, we examined the situation in which customers arrive and are processed at constant rates. While there are many applications for which that "model" is acceptable, there are times when it is not acceptable. For example, customers do not arrive at a fast food restaurant at a rate of exactly one each minute, nor does it take the same amount of time to serve each customer. A comparison of the types of arrival rates for the situation portrayed in the last section and in this section is shown in Figure 1.

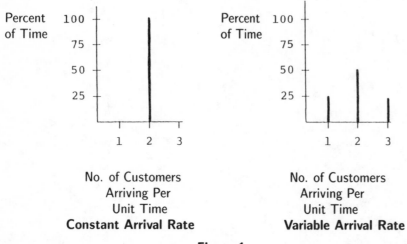

No. of Customers
Arriving Per
Unit Time
Constant Arrival Rate

No. of Customers
Arriving Per
Unit Time
Variable Arrival Rate

Figure 1

For the variable arrival rate situation, we do not know exactly how many customers will arrive during any unit of time. We only know that 25% of the time, 1 customer arrives in a unit of time; 50% of the time, 2 customers arrive per time unit; and 25% of the time, 3 customers arrive per time unit. It is usually assumed that the number arriving in one time unit does not depend on the number of arrivals in previous or subsequent time intervals. A similar situation exists in this section for processing or service times in that the required service time per customer can vary.

Consider the service time situations shown in Figure 2 and answer the following questions.

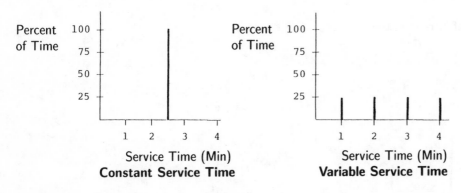

Figure 2

50. For the constant service time situation:

    a. What percent of the time does it require 3 minutes to service a customer? _____

    b. What percent of the time does it require 2.5 minutes to service a customer? _____

    c. What is the average service time? _____

51. For the variable service time situation:

    a. What percent of the time does it require 4 minutes to service a customer? _____

    b. 4.5 minutes? _____

    c. 1.0 minute? _____

    d. What is the average service time? _____

    Thus, we see that two different situations can have the same average service times.

    Simulation is a technique which can be used to mimic (model) real situations which include uncertainties. The simulation itself is a procedure which behaves like the real process. For example, the time lines of the previous sections were simulations of the spraying machine process.

    Let's see how we might construct time lines for variable arrival times and service times.

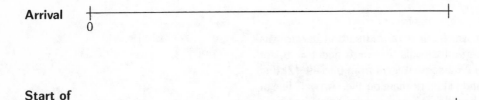

We begin by selecting the number of arrivals at time = 0 according to the frequency data in Figure 3.

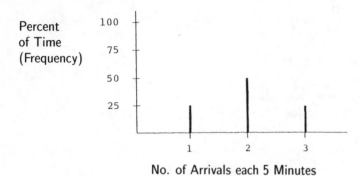

No. of Arrivals each 5 Minutes

Figure 3

In this example, we shall assume that all of the customers for a given five minute interval arrive at the beginning of the interval.

Now, how can we choose a number of arrivals to write on our time line? Clearly, we want a way which is twice as likely to choose 2 arrivals as either 1 or 3. One way to simulate the number of arrivals at the beginning of a five-minute interval would be to choose at random an integer from 0 to 99. The phrase "at random" means every integer has an equal chance of being selected. We could pretend (simulate) one arrival, if any integer 0 to 24 was chosen, two arrivals, if any integer 25 to 74 was chosen, and three arrivals if any of the 25 integers 75 to 99 was chosen. Clearly, this approach does fulfill our need to select two arrivals twice as frequently as either 1 or 3.

52. What number of arrivals would correspond to each of the following random numbers:

    a. 16? _____

    b. 25? _____

    c. 96? _____

    d. 47? _____

53. In fact, if the numbers are drawn at random, what percent of the time will:

    a. 1 arrival every 5 min occur? _____

    b. 2 arrivals every 5 min occur? _____

    c. 3 arrivals every 5 min occur? _____

But these are exactly the frequencies at which our data told us that our arrivals should occur. Thus, to find the numbers of arrivals at $t = 0$ and $t = 5$, for example, all we need to do is: (1) obtain 2 random integers from 0 to 99; (2) find the corresponding numbers of arrivals, and (3) mark them on the "Arrival" line of our chart at the appropriate times. But how do we select numbers "at random"? In fact, people have spent a great deal of time and effort developing computer programs and processes for producing random numbers. These programs have been used to prepare tables of random numbers for use in "simulations" like ours. Table 1 contains such random numbers.

In Table 1, all of the numbers 0, 1, ..., 9 appear with almost the same frequency. By combining numbers in pairs, we obtain the 100 numbers 00 to 99. In order to obtain or draw the needed random numbers, the table should be entered in a random manner by first selecting a page of the table at random, and then selecting, at random, a "starting location" on that page. One way this can be done is to close your eyes and place a finger on the page. The "starting number" is the two-digit number under your finger. Since the digits are thoroughly mixed in the table, the entire sample of random numbers needed may be taken as successive horizontal numbers in the table once a starting location has been determined randomly.

For example, let us use the random number table to identify the number of arrivals at $t = 0$ and $t = 5$ minutes. If we choose at random page 2 of Table 1, and randomly place a finger down, suppose it falls on the 83, in the 6th row and 11th column of 2 digit numbers. The 83 indicates that 3 arrivals will occur at time $= 0$. The next random number, 64, indicates that at time $= 5$ minutes, 2 arrivals will occur. These arrivals can be noted on the time line. The process could be continued to find the numbers of arrivals occurring at the beginning of succeeding five minute intervals. It can also be noted that c_1 begins service at $t = 0$. But how long is c_1's service time?

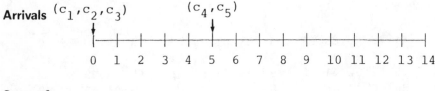

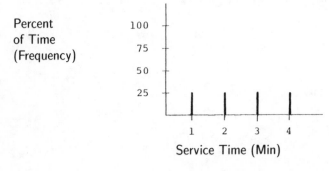

We now apply the same concept to determine service times for each of our five customers. The original service time chart appears again in Figure 4.

Percent
of Time
(Frequency)

Service Time (Min)

Figure 4

Table 1 (Page 1)

```
27 89 70 31 42     52 03 35 60 36     82 05 17 29 06     98 22 45 48 75     52 16 21 23 60
59 93 94 48 05     64 89 47 42 96     24 80 52 40 37     20 63 61 04 02     00 82 29 16 65
08 42 26 89 53     19 64 50 93 03     23 20 90 25 60     15 95 33 47 64     35 08 03 36 06
99 01 90 25 29     09 37 67 07 15     38 31 13 11 65     88 67 67 43 97     04 43 62 76 59
12 80 79 99 70     80 15 73 61 47     64 03 23 66 53     98 95 11 68 77     12 17 17 68 33


66 06 57 47 17     34 07 27 68 50     36 69 73 61 70     65 81 33 98 85     11 19 92 91 70
31 06 01 08 05     45 57 18 24 06     35 30 34 26 14     86 79 90 74 39     23 40 30 97 32
85 26 97 76 02     02 05 16 56 92     68 66 57 48 18     73 05 38 52 47     87 18 19 91 92
16 78 74 80 93     83 40 59 75 27     66 65 52 22 52     59 60 23 29 49     07 82 72 09 32
32 83 36 86 75     48 59 24 05 07     00 45 28 60 37     75 72 76 01 55     82 74 16 18 70


52 33 26 64 01     72 06 57 09 61     46 26 87 73 47     43 53 30 17 59     83 09 98 95 66
80 55 95 90 68     36 92 21 91 98     96 39 58 47 11     69 14 62 78 26     78 15 55 75 87
30 73 21 62 88     08 78 73 95 16     05 92 91 22 30     49 03 14 72 87     71 73 34 39 28
30 41 49 11 28     08 56 09 06 53     63 64 39 70 95     38 92 81 24 52     76 09 94 47 96
00 58 46 79 93     93 38 18 85 32     23 70 21 17 59     16 49 44 19 38     54 60 16 25 08


83 87 83 76 16     08 73 43 25 38     41 45 60 83 32     59 83 01 24 14     13 49 20 36 80
71 26 80 95 10     04 06 96 38 27     07 74 20 15 12     33 87 25 01 62     52 98 94 62 46
11 71 39 64 16     94 57 91 33 92     25 02 92 61 38     97 19 11 94 75     62 03 19 32 42
05 04 98 88 46     62 09 06 83 05     36 56 14 66 35     63 46 71 43 00     49 09 19 81 80
57 07 77 51 30     38 20 86 83 92     99 01 68 41 48     27 74 51 80 81     39 80 72 89 35


55 07 94 55 99     36 04 98 62 67     93 15 21 04 38     92 41 47 02 06     31 51 39 93 00
18 83 39 37 57     80 43 07 35 21     38 95 35 43 53     77 53 19 82 05     23 56 41 93 19
06 74 20 52 97     19 14 63 80 17     96 59 12 90 08     18 49 05 13 42     64 47 31 07 83
05 26 62 91 29     14 17 72 98 49     89 59 43 00 95     41 80 11 95 06     07 06 03 39 10
82 83 85 09 80     06 59 07 52 63     27 52 46 56 88     63 87 34 55 79     41 29 76 21 97
```

Table 1 (Page 2)

```
19 61 39 64 14    27 84 17 75 62    03 90 68 74 30    00 04 15 55 19    18 57 78 00 70
11 28 66 19 00    08 56 74 06 47    70 42 71 44 09    48 12 28 38 36    29 39 40 98 50
77 73 60 36 25    03 29 46 87 56    45 71 95 17 69    64 94 06 83 93    81 84 78 47 81
35 86 31 86 81    34 03 06 27 03    26 65 62 44 45    36 90 32 34 65    23 53 84 33 11
21 30 01 66 66    45 27 59 33 67    07 11 56 78 70    89 16 25 55 07    33 37 56 71 18


03 98 88 37 59    46 62 98 06 91    83 64 67 68 09    82 59 79 60 59    37 74 83 24 19
06 53 83 05 47    63 64 33 30 36    56 75 52 16 39    54 94 04 41 83    90 20 77 22 72
14 53 66 35 63    43 98 51 37 63    46 08 43 58 71    56 80 74 63 81    94 87 56 95 73
71 03 43 00 74    58 49 27 81 49    09 28 82 82 35    18 95 69 22 67    90 58 97 68 59
19 81 19 49 36    80 57 04 32 38    44 83 54 60 07    88 39 10 90 17    48 51 42 38 91


24 35 89 58 72    14 27 14 02 85    30 65 15 41 82    07 67 34 56 10    56 10 74 70 29
70 95 77 43 59    38 92 64 62 54    39 00 90 17 (73)  74 28 77 52 51    65 34 46 74 15
21 81 85 93 13    93 27 88 17 57    05 68 67 31 56    07 08 28 50 46    31 85 33 77 74
70 34 20 79 26    72 06 70 40 26    75 17 97 88 23    33 61 53 03 18    82 68 88 44 21
91 46 62 36 83    90 92 37 36 12    75 52 93 34 53    84 43 75 48 62    13 11 05 74 36


21 72 44 58 51    71 83 47 37 27    93 64 94 21 68    74 05 50 00 27    14 99 51 66 48
24 52 83 19 22    76 09 16 59 32    41 23 05 08 94    89 85 74 48 11    03 77 86 95 36
14 62 19 97 30    13 26 78 89 62    68 84 01 74 96    75 30 36 81 32    53 60 90 06 24
70 29 88 36 74    52 21 30 43 80    02 31 87 70 94    53 67 77 65 62    00 89 11 01 43
03 33 82 91 14    88 99 87 16 74    43 10 43 43 96    69 98 36 20 97    07 36 18 67 86


05 13 18 00 65    47 50 56 16 57    63 25 62 95 57    62 81 12 46 20    76 16 45 95 47
47 07 96 00 89    58 81 22 29 58    46 07 22 52 58    05 40 40 26 85    03 89 69 90 79
81 05 47 31 35    51 25 60 07 15    34 39 84 22 32    38 52 99 95 83    19 59 70 90 55
37 14 90 87 52    34 90 88 27 54    71 64 58 85 88    30 18 40 66 75    60 34 91 17 33
24 60 09 07 21    85 62 15 53 28    35 04 94 62 92    51 69 12 76 98    68 78 96 07 36
```

Table 1 (Page 3)

```
66 09 23 28 22    39 05 95 28 49    58 44 42 03 49    78 52 19 74 06    70 79 10 01 76
04 79 93 13 97    43 93 38 62 23    84 35 78 70 20    30 18 24 63 43    36 45 21 87 55
92 83 35 42 94    15 91 84 35 92    09 82 90 63 21    68 61 46 95 95    32 65 46 10 24
07 85 69 02 79    96 55 18 72 09    57 33 14 44 41    43 54 03 49 66    10 13 26 46 62
06 51 12 94 99    19 91 78 23 23    54 35 16 81 45    19 35 17 77 00    99 01 46 41 46


93 92 96 77 34    07 96 54 45 33    52 38 02 42 01    11 79 51 12 30    46 36 23 94 09
55 33 85 32 93    24 36 23 70 23    49 40 53 16 80    70 21 19 38 17    91 48 57 96 05
36 55 75 47 70    42 35 41 46 69    74 08 40 03 06    14 33 92 81 73    38 28 42 80 18
58 17 86 23 58    38 43 54 29 81    52 39 44 40 43    23 07 39 61 77    65 77 21 68 92
52 15 52 45 08    10 45 67 37 89    05 05 30 08 74    53 66 66 23 86    72 22 58 32 25


21 02 15 71 88    09 92 64 79 42    60 22 61 08 52    78 35 33 22 81    19 08 39 77 45
24 38 17 59 29    28 14 16 49 55    40 81 90 02 41    44 44 55 03 60    24 16 04 36 46
32 72 61 30 78    14 57 60 81 04    09 88 98 01 28    07 90 99 73 32    94 29 15 04 18
43 49 92 18 52    53 32 10 95 33    90 12 83 69 33    21 93 86 66 61    99 68 48 54 46
65 08 88 26 58    16 85 65 13 61    71 00 98 92 79    65 93 17 85 48    38 34 10 63 48


81 51 05 56 24    92 46 96 98 29    39 93 29 93 63    38 52 56 03 31    80 21 62 76 43
69 50 19 71 38    54 79 59 40 38    19 60 16 01 63    57 82 74 41 40    31 55 72 08 01
39 69 14 73 19    75 05 86 66 61    81 62 80 44 48    39 67 39 59 84    39 58 63 86 85
81 14 43 46 70    86 71 42 17 22    63 54 76 11 17    54 35 58 51 47    02 83 86 45 50
78 93 66 69 22    14 28 46 99 66    55 27 32 26 18    47 26 24 31 10    65 91 91 57 26


47 78 49 49 89    55 32 11 15 32    08 30 42 10 31    12 36 45 70 41    75 03 47 12 63
10 25 08 95 59    15 82 66 98 63    40 99 74 47 42    07 40 41 61 57    03 60 64 11 45
86 60 90 85 06    46 18 80 62 05    17 90 11 43 63    80 72 50 27 39    31 13 41 79 48
68 61 24 78 18    96 83 55 41 18    56 67 77 53 59    98 92 41 39 68    05 04 90 67 00
82 89 40 90 20    50 69 95 08 30    67 83 28 10 25    78 16 25 80 72    42 60 19 04 37
```

If we continue on from the last number selected in the Random Number Table and draw the next five two-digit random numbers, we obtain 67, 68, 09, 82, 59.

54. **a.** 67 leads to a service time of _____ min for customer _____.

b. 68 leads to a service time of _____ min for customer _____.

c. 09 leads to a service time of _____ min for customer _____.

d. 82 leads to a service time of _____ min for customer _____.

e. 59 leads to a service time of _____ min for customer _____.

55. With this information in hand, fill in the remaining information on the time lines for the servicing and departure of these five customers, just as we did in the last section.

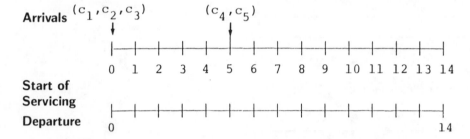

In fact, we can perform the calculations (e.g., average waiting time, time until first no-wait customer, etc.) as in the last section, once the numbers of customers and their times of arrival and service times are established. If we repeated the process with new random numbers, the results would be different. Therefore, if we want to get a true evaluation of such a variable rate process, the results of many such calculations should be reviewed and analyzed. For example, the waiting times should be computed for many repeats (called replications) of the situation and reviewed to see the average, maximum, and minimum values that it takes on. The benefit of using a computer for making such repetitive calculations should be apparent.

Now that you have the idea of how the variable arrival rate and service time problem differs from constant rate problems, we will apply the method to a real-life queueing problem, that of a post office. However, some differences will be obvious. We will no longer use unwieldy time lines to keep track of data. Tables will be used instead. Multiple servers will also be present, which will complicate the bookkeeping but not change the basic concepts you have just learned.

Post Office Problem

A small-town post office in Maryland has the capacity of providing five clerks to work simultaneously. There is not enough work space for any additional clerks. Customers enter a common queue and are serviced by the first available clerk.

During the 1981 "Christmas Rush" season, data on arrival times and service times were collected. Table 2 contains data on arrivals. The first row in the table is the number of arrivals per minute, i. The second row records the fraction of occurrences of i, denoted by F_i. That is,

$$F_i = \frac{\text{number of minutes sampled for which the arrivals per minutes is } i}{\text{total number of minutes sampled}}.$$

Table 2
Arrival Times

Number of arrivals per minute	i	0	1	2	3	4	5	6	7
Fraction of occurrences of i	F_i	0.07	0.23	0.10	0.30	0.07	0.10	0.06	0.07

This table is interpreted in the following way: For 7% of the one-minute intervals sampled, no customers arrived. For 23% of the one-minute intervals sampled, exactly one customer arrived, and so on, ..., while finally for 7% of the one-minute intervals sampled, exactly 7 customers arrived.

56. For which value of i is F_i largest? _____

Smallest? _____

57. Compute

$$\sum_{i=0}^{7} F_i = F_0 + F_1 + F_2 + F_3 + F_4 + F_5 + F_6 + F_7.$$

58. Why should you have anticipated the answer *without* doing the addition?

59. Compute λ, the average number of arrivals per minute.

$$\lambda = \sum_{i=0}^{7} i \cdot F_i = 0 \cdot F_0 + 1 \cdot F_1 + 2 \cdot F_2 + \cdots + 7 \cdot F_7.$$

Table 3 contains data on customer service times. The first row gives service times, j (to the nearest 10 seconds). The second row records the fraction

$$G_j = \frac{\text{number of samples for which the service time was } j}{\text{total number of samples}}.$$

Table 3
Service Times

Service times (Seconds)	j	10	20	30	40	50	60	70	80	90	100	110
Fraction of occurrences of j	G_j	0.15	0.15	0.11	0.02	0.08	0.10	0.08	0.03	0.03	0.05	0.05

Service times (Seconds)	j	120	130	140	150	160	170	180	190	200	210	220
Fraction of occurrences of j	G_j	0.05	0.02	0.00	0.02	0.03	0.00	0.00	0.00	0.00	0.00	0.03

60. For which j is G_j the largest? _____
Smallest? _____

61. Compute: $G_{10} + G_{20} + G_{30} + \cdots + G_{220} =$ _____
Again, why should you have anticipated the answer? _____
_____

62. Compute, μ, the average service time:

$$\mu = 10 \cdot G_{10} + 20 \cdot G_{20} + 30 \cdot G_{30} + \cdots + 220 \cdot G_{220}$$

_____

With this information, we will be using the simulation approach to gain insights as to how many servers the post office should provide in order to keep the waiting time for customers within reasonable limits.

Application of Random Number Procedure to Simulating Numbers of Arrivals

We can use the random number table to simulate the number of arrivals for each minute of a 30-minute period. From Table 2, repeated here,

Table 2

No. of Arrivals Per Minute	i	0	1	2	3	4	5	6	7
Fractions of Occurrences of i	F_i	0.07	0.23	0.10	0.30	0.07	0.10	0.06	0.07

we construct an assignment scheme as follows.
During the first minute there are

> 0 arrivals if any one of the numbers 00 to 06 is drawn,
> 1 arrival if any one of the numbers 07 to 29 is drawn,
> 2 arrivals if any one of the numbers 30 to 39 is drawn,
> 3 arrivals if any one of the numbers 40 to 69 is drawn,
> 4 arrivals if any one of the numbers 70 to 76 is drawn,
> 5 arrivals if any one of the numbers 77 to 86 is drawn,
> 6 arrivals if any one of the numbers 87 to 92 is drawn,
> 7 arrivals if any one of the numbers 93 to 99 is drawn.

Suppose the starting location in the random number table is p. 2 of Table 1, row 12 under the 15th column of 2-digit numbers. The corresponding number, 73, will be used to simulate a number of arrivals during the first minute.

63. Using the graph above, and the random number 73, how many arrivals occur in the first minute? _____

64. Continuing along the same row of Table 1, the next two random numbers are _____ and _____ .

65. Corresponding to each of these we would assign how many arrivals to the second and third minutes of the simulation?
_____ and _____

As the entries in one row of random numbers are used up, you proceed to the next row and continue.

It was by this procedure that the numbers of arrivals for each of 30 simulated post office minutes were determined. The results are shown in Table 4. (The benefits of a computer's speed are obvious in a situation like this.)

Table 4

Use of random number table to simulate the number of arrivals per minute for each minute of a 30-minute period

Minute of Simulation	Random Number[1]	Arrivals/Minute
1	73	4
2	74	4
3	28	1
4	77	5
5	52	3
6	51	3
7	65	3
8	34	2
9	46	3
10	74	4
11	15	1
12	21	1
13	81	5
14	85	5
15	93	7
16	13	1
17	93	7
18	27	1
19	88	6
20	17	1
21	57	3
22	05	0
23	68	3
24	67	3
25	31	2
26	56	3
27	07	1
28	08	1
29	28	1
30	50	3

66. What random number was associated with the number of arrivals in the 15th minute of the simulation? _____

67. Why is 7 arrivals/minute assigned to this random number?

_____

_____

---

[1] Based on starting location in page 2 of Table 1, row 12, column 15.

68. Add 4 minutes of simulation (minutes 31 to 34) to Table 4, continuing in the random number table where Table 4 ended.

	Minute	Random Number	Arrivals/Minute
a.	31	_____	_____
b.	32	_____	_____
c.	33	_____	_____
d.	34	_____	_____

In the next section, we will generate service times for each of our customers.

Application of Random Numbers Procedure to Simulating Service Times
The first two columns of Table 5 are the same as Table 3, which you may recall gave the frequencies with which different service times occurred. The third column of Table 5 lists the random numbers that would be assigned to service time j if our random procedure were used for simulating service times. For example, a random number of 90 would be assigned to a service time of 103 seconds.

Table 5

**Service times and associated ranges
of random numbers**

Service Times j	Fraction of Occurrences of j G_j	Random numbers to be assigned to j
10	0.15	00 to 14
20	0.15	15 to 29
30	0.11	30 to 40
40	0.02	41 and 42
50	0.08	43 to 50
60	0.10	51 to 60
70	0.08	61 to 68
80	0.03	69 to 71
90	0.03	72 to 74
100	0.05	75 to 79
110	0.05	80 to 84
120	0.05	85 to 89
130	0.02	90 and 91
140	0.00	None
150	0.02	92 and 93
160	0.03	94 to 96
170	0.00	None
180	0.00	None
190	0.00	None
200	0.00	None
210	0.00	None
220	0.03	97 to 99

Table 6

Simulated service times for 87 consecutive customers

Customer	Random Number	Service Time	Customer	Random Number	Service Time
1	82	110	45	68	70
2	66	70	46	61	70
3	98	220	47	24	20
4	_____	_____	48	78	100
5	_____	_____	49	18	20
6	_____	_____	50	96	160
7	_____	_____	51	83	110
8	47	50	52	55	60
9	42	40	53	41	40
10	07	10	54	18	20
11	40	30	55	56	60
12	41	40	56	67	70
13	61	70	57	77	100
14	57	60	58	53	60
15	03	10	59	59	60
16	60	60	60	98	220
17	64	70	61	92	150
18	11	10	62	41	40
19	45	50	63	39	30
20	86	120	64	68	70
21	60	60	65	05	10
22	90	130	66	04	10
23	85	120	67	90	130
24	06	10	68	67	70
25	46	50	69	00	10
26	18	20	70	82	110
27	80	110	71	89	120
28	62	70	72	40	30
29	05	10	73	90	130
30	17	20	74	20	20
31	90	130	75	50	50
32	11	10	76	69	80
33	43	50	77	95	160
34	63	70	78	08	10
35	80	110	79	30	30
36	72	90	80	67	70
37	50	50	81	83	110
38	27	20	82	28	20
39	39	30	83	10	10
40	31	30	84	25	20
41	13	10	85	78	100
42	41	40	86	16	20
43	79	100	87	25	20
44	48	50			

Table 6 contains a service time for each of the 87 customers who arrived during the 30 simulated minutes of post office operations. These times were generated by drawing random numbers, starting at the following location: Table 1, page 3, row 22, column 7, and assigning service times according to Table 5.

69. Fill in the random numbers and service times for customers 4, 5, 6, and 7 in Table 6. Space has been left in the table.

70. Add 4 customers (customer 88 to 91) to Table 6, continuing in the random number table where Table 6 left off.

	Customer	Random Number	Service Time
a.	88	_____	_____
b.	89	_____	_____
c.	90	_____	_____
d.	91	_____	_____

Complete Simulation

We have now defined the "workload" (e.g., the number of customers and the service time for each) which the post office will face during *one* possible 30 minute interval. Now we can study the capability of the post office to process the workload based upon the number of clerks available. Let's start by assuming there are 2 clerks, *A* and *B*. Table 7 gives "simulation" results for our 87 consecutive customers.

The results in Table 7 are based on data from Table 4, which gives the number of arrivals, and Table 6, which gives service times. To see this, let's examine the first several rows of Table 7.

(a) Customer 1 arrives at the start of the simulation. Clerk *A* is available to serve customer 1, so there is no waiting time. The service time is 110 seconds (see Table 6), so customer 1 will be finished at 110 seconds.

(b) Customer 2 enters at time $= 0$ and is serviced immediately by Clerk *B*. Customer 2 requires 70 seconds to service (see Table 6), and he finishes at the 70 second mark.

(c) The third customer also arrives at time $= 0$ and from Table 6, we see that that customer requires 220 seconds of service time. But Customer 3 must wait until a clerk is available. Clerk *B* becomes available at time $= 70$ and immediately starts to process Customer 3. Clerk *B* starts serving Customer 3 at time $t = 70$ and continues for 220 seconds until time $t = 290$. The remaining customers are treated in the same fashion.

Table 7

Complete Simulation

Customer	Arrival Time (seconds after start of simulation)	Waiting Time	Start of Service	Service Time	Finish of Service	Clerk
1	0	0	0	110	110	A
2	0	0	0	70	70	B
3	0	70	70	220	290	B
4	0	110	110	70	180	A

Table 7 (cont.)
Complete Simulation

Customer	Arrival Time (seconds after start of simulation)	Waiting Time	Start of Service	Service Time	Finish of Service	Clerk
5	60	120	180	30	210	A
6	60	150	210	220	430	A
7	60	230	290	90	380	B
8	60	320	380	50	430	B
9	120	310	430	40	470	A[3]
10	180	250	430	10	440	B
11	___	___	___	___	___	___
12	180	290	470	40	510	A
13	180	290	470	70	540	B
14	180	330	510	60	570	A
15	___	___	___	___	___	___
16	240	310	550	60	610	B
17	240	330	570	70	640	A
18	300	310	610	10	620	B
19	300	320	620	50	670	B
20	300	340	640	120	760	A
21	360	310	670	60	730	B
22	360	370	730	130	860	B
23	360	400	760	120	880	A
24	420	440	860	10	870	B
25	420	450	870	50	920	B
26	480	400	880	20	900	A
27	480	420	900	110	1010	A
28	480	440	920	70	990	B
29	540	450	990	10	1000	B
30	540	460	1000	20	1020	B
31	540	470	1010	130	1140	A
32	540	480	1020	10	1030	B
33	600	430	1030	50	1080	B
34	660	420	1080	70	1150	B
35	720	420	1140	110	1250	A
36	720	430	1150	90	1240	B
37	720	520	1240	50	1290	B
38	720	530	1250	20	1270	A
39	720	550	1270	30	1300	A
40	780	510	1290	30	1320	B
41	780	520	1300	10	1310	A
42	780	530	1310	40	1350	A
43	780	540	1320	100	1420	B
44	780	570	1350	50	1400	A
45	840	560	1400	70	1470	A

---

[3]If 2 servers are available at the same time, it does not matter which is chosen by the next customer.

Table 7 (cont.)

Complete Simulation

Customer	Arrival Time (seconds after start of simulation)	Waiting Time	Start of Service	Service Time	Finish of Service	Clerk
46	840	580	1420	70	1490	B
47	840	630	1470	20	1490	A
48	840	650	1490	100	1590	B
49	840	650	1490	20	1510	A
50	840	670	1510	160	1670	A
51	840	750	1590	110	1700	B
52	900	770	1670	60	1730	A
53	960	740	1700	40	1740	B
54	960	770	1730	20	1750	A
55	960	780	1740	60	1800	B
56	960	790	1750	70	1820	A
57	960	840	1800	100	1900	B
58	960	860	1820	60	1880	A
59	960	920	1880	60	1940	A
60	1020	880	1900	220	2120	B
61	1080	860	1940	150	2090	A
62	1080	1010	2090	40	2130	A
63	1080	1040	2120	30	2150	B
64	1080	1050	2130	70	2200	A
65	1080	1070	2150	10	2160	B
66	1080	1080	2160	10	2170	B
67	1140	1030	2170	130	2300	B
68	1200	1000	2200	70	2270	A
69	1200	1070	2270	10	2280	A
70	1200	1080	2280	110	2390	A
71	1320	980	2300	120	2420	B
72	1320	1070	2390	30	2420	A
73	1320	1100	2420	130	2550	B
74	1380	1040	2420	20	2440	A
75	1380	1060	2440	50	2490	A
76	1380	1110	2490	80	2570	A
77	1440	1110	2550	160	2710	B
78	1440	1130	2570	10	2580	A
79	1500	1080	2580	30	2610	A
80	1500	1110	2610	70	2680	A
81	1500	1180	2680	110	2790	A
82	1560	1150	2710	20	2730	B
83	1620	1110	2730	10	2740	B
84	1680	1060	2740	20	2760	B
85	1740	1020	2760	100	2860	B
86	1740	1050	2790	20	2810	A
87	1740	1070	2810	20	2830	A

From Table 7, determine:

71. All table entries for customers 11 and 15. (Place your answers directly in the blanks left in Table 7.)

72. The waiting time for customer 14. _____

73. The average waiting time for

    a. customers 1 to 5 _____

    b. customers 6 to 10 _____

    c. customers 11 to 15 _____

    d. customers 16 to 20 _____

    e. customers 41 to 45 _____

    f. customers 83 to 87 _____

The answers to question 72 indicate how quickly the waiting time can grow when the post office is understaffed. Within 7 minutes of simulation, the waiting time grows to over 7 minutes. If the post office remains understaffed for 30 minutes, the waiting time can grow to 1070 seconds or almost 18 minutes for customer 87.

74. Use the data from Tables 4 and 6 to determine the waiting times for customers 1 to 15 when 3 clerks are available. (Use the following table for your answers. Some have been filled in for you.)

Customer	Arrival Time	Waiting Time	Start of Service	Service Time	Finish of Service	Clerk
1	0	0	0	110	110	A
2	0	0	0	70	70	B
3	0	0	0	220	220	C
4	0	70	70	70	140	B
5	60	___	___	___	___	___
6	60	___	___	___	___	___
7	60	___	___	___	___	___
8	60	___	___	___	___	___
9	120	___	___	___	___	___
10	180	___	___	___	___	___
11	180	___	___	___	___	___
12	180	___	___	___	___	___
13	180	___	___	___	___	___
14	180	___	___	___	___	___
15	240	___	___	___	___	___

75. Based on the results of question 74, determine the average waiting time for

    a. customers 1 to 5 _____

    b. customers 6 to 10 _____

    c. customers 11 to 15 _____

    Comparing the results of question 73 and 75, we see that the waiting times drop substantially when 3 clerks are available.

76. Use the data from Tables 4 and 6 to determine the waiting times for customers 1 to 15 when 4 clerks are available. (Use the following table for your answers. Some have been filled in for you.)

Customer	Arrival Time	Waiting Time	Start of Service	Service Time	Finish of Service	Clerk
1	0	0	0	110	110	A
2	0	0	0	70	70	B
3	0	0	0	220	220	C
4	0	0	0	70	70	D
5	60	___	___	30	___	___
6	60	___	___	220	___	___
7	60	___	___	90	___	___
8	60	___	___	50	___	___
9	120	___	___	40	___	___
10	180	___	___	10	___	___
11	180	___	___	30	___	___
12	180	___	___	40	___	___
13	180	___	___	70	___	___
14	180	___	___	60	___	___
15	240	___	___	10	___	___

77. Based on the results of question 76, determine the average waiting time for

    a. customers 1 to 5 _____

    b. customers 6 to 10 _____

    c. customers 11 to 15 _____

    You were probably not surprised to see another big drop in waiting times when 4 clerks are available. However, you may have been surprised to see that some customers still have to wait 30 seconds. Recall that the average service time was 1 minute and the average number of arrivals/minute was 2.96. So, it would seem that 4 clerks would be sufficient to reduce the waiting times to zero. Not so!! Because of the randomness of arrival and service times, queues can still develop.

For the past several pages, you have worked on a *simplified*, but realistic simulation and should be gaining insights into the questions that such an analysis can answer. Consider the following points which would complicate the simulation but make it more realistic. First, we assumed that each customer arrived at the *beginning* of every 1-minute interval. Clearly, that's not realistic, How could you use the random number procedure to assign arrival times throughout the 1-minute interval? Also, we assumed that the number of arrivals in any interval didn't depend on the numbers of arrivals from previous periods. But most post-offices have busy periods. During such times, large numbers of arrivals are likely in successive periods. How could we alter the simulation approach to reflect busy and slack periods realistically? Finally, how could the simulation be changed to reflect "express lines" being set up for customers requiring only short servicing like buying stamps?

Congratulations if you've had the perseverance to work through the examples. Such work is tedious, but aids in understanding the process and is essential if you want to program and computerize such calculations.

B. Simulation in Practice

Repeated Runs

In our post office problem, we restricted the analysis to one "run" (sometimes called replication) of the simulation process. This one run consisted of selecting arrival and service times and computing waiting times. If we were to repeat the process, that is, do a second replication entering the random number tables at new random starting locations, no doubt we would generate a different set of arrival, service, and waiting times.

When simulation is used in practical problems, the user is never satisfied with one run. Often 100 or 1,000 or even more replications are made depending on the complexity of the model or the importance of the problem. We are more comfortable having a collection of results from many replications. We can then study the variability of the results and estimate quantities, such as the probability that the average waiting time exceeds a certain value. Obviously, a computer is the perfect tool for making so many repetitive calculations.

C. Wide Applicability of Simulation

Many important mathematical problems are so big or complex that simulation techniques provide the only hope for solutions. Most applications of simulation models involve random phenomena. For example, in models of queues, the random variables were arrival and service times; in inventory models the variables include customer demand and delivery times; in marketing models the variables include possible new product discoveries. Other large industrial and military models may contain more than 15 random variables, each with its own probability distribution (similar to Tables 1 and 2). Frequently, such simulations require the generation of thousands, sometimes millions of random numbers.

Although the formulation of these models is detailed and time-consuming, the solutions, in concept, are no more difficult than our sample waiting line problem.

Summary

You have completed an introduction to the study of queues or waiting lines. Through examples, you probably realize that you find yourself in queues almost every day.

There is a wide spectrum of queue complexity, from single server systems with constant arrival and service rates, to multiple servers perhaps each having random and distinct service and arrival times. The goal of the study of such systems is often to see if it is possible to strike a balance between waiting time and cost by providing enough servers so that waiting times are reduced to customer-acceptable level.

As we have seen in the text, one method of analyzing waiting lines is to use simulation. Data are collected and a computer program is frequently used to simulate the activity of interest. Special languages have been developed especially for queuing simulations: SIMSCRIPT and GPSS (General Purpose Simulation System) are two popular examples. The interested student could learn one or both of these languages or develop simulations using FORTRAN or BASIC. In any case, the applications of queueing and simulation are many and useful, and you are limited only by your own initiative and ingenuity!

Special Projects

1. For part III, write a computer program to determine

    (a) the actual service time when the first "no-wait" customer arrives.

    (b) the available service time when the first "no-wait" customer arrives.

    (c) the rank order of the first "no-wait" customer.

2. Write a computer program to simulate waiting lines experienced in Maryland as described in part IV. Use your program to run 10 replications when 2 clerks are available.

3. Do an analysis of the waiting lines at your post office or busiest traffic intersection.

References

Alford et al., 1973. *Man and His Technology (Chapter 3)*. McGraw-Hill Book Company, New York, NY.

Cooper et al., 1977. *Introduction to Operations Research Models*. W. B. Saunders Co., Philadelphia, Pennsylvania.

Joseph Panico, 1972. *Queuing Theory*. Prentice-Hall, Inc., Englewood Cliffs, New Jersey.

Saaty, 1961. *Queueing Theory*. McGraw-Hill Book Company, New York, NY.

Answers

1. Possible answers include: computer programs waiting for a time sharing system; waiting in fast food service lines; barbers shop; waiting in line for class pictures; buying tickets for basketball championship games.

2. 58.33%

3. 0

4. 1 arrival each 7 seconds.

5. 100%

6. **a.** The queue will continue to grow.
   b. No queue will form.
   c. No queue will form.

7. **a.** Add another processor.
   b. Reduce the arrival rate of items to be serviced.

8. **a.** Cans being labeled in a canning factory
   b. Soft drink bottles being placed into cartons

9. **a.** 15
   b. 50
   c. 120

10.
Time Factor	Waiting Line
Machine B begins	120
End of one hour	70
End of two hours	20
End of three hours	0

11.
Time Factor	Waiting Line
Machine B begins	210
End of one hour	160
End of two hours	110
End of three hours	60
End of four hours	10
End of five hours	0

Answer to original question is <u>5</u>.

12.
Time Factor	Waiting Line
Machine B begins	$30t$ or $30t$
End of one hour	$30(t+1) - 80$ or $30t - 50$
End of two hours	$30(t+2) - 2 \cdot 80$ or $30t - 2 \cdot 50$
End of three hours	$30(t+3) - 3 \cdot 80$ or $30t - 3 \cdot 50$
End of four hours	$30(t+4) - 4 \cdot 80$ or $30t - 4 \cdot 50$
.	. .
.	. .
.	. .
End of M hours	$30(t+M) - M \cdot 80$ or $30t - M \cdot 50$

13. The smallest integer M for which

$$M \geq \tfrac{3}{5}t.$$

14. One must assume that $x \leq y$ so that the queue will not simply continue to grow beyond bounds.

15.
Time Factor	Waiting Line
Machine B begins	xt or xt
End of one hour	$x(t+1) - y$ or $xt - (y-x)$
End of two hours	$x(t+2) - 2y$ or $xt - 2(y-x)$
End of three hours	$x(t+3) - 3y$ or $xt - 3(y-x)$
End of four hours	$x(t+4) - 4y$ or $xt - 4(y-x)$
.	. .
.	. .
.	. .
End of M hours	$x(t+M) - My$ or $xt - M(y-x)$

16. **a.** The smallest integer M for which

$$\frac{xt}{y-x} \leq M.$$

   b. It is necessary that $x < y$. Otherwise the denominator of the above fraction would be negative, a physically impossible result. If $x = y$, there is no change in the queue.

17. 20

18. 10

19. 21

20. 20

21.

Customer	Waiting Time
e_1	0
e_2	2
e_3	4
e_4	6
e_5	8
e_6	10
c_1	11
c_2	8
c_3	5
c_4	2

22. 5.6 minutes

23. 0 minutes

24. The seventh

25. 11

26. 44

27. 44

28. 9.36 minutes

29. 44

30. 34

31. $24S$

32. $(w + 6)S$

33. $(w + n)S$

34. 32

35. 44

36. $F + 56$

37. $F + (9 \cdot T)$

38. $F + (n \cdot T)$

39. Actual Service Time $\leq$ Available Operating Time.

40. 20

41. $S < T$

42. $(W + n) \cdot S$

43. $nT + F$

44. $(W + n) \cdot S \leq nT + F$

45. $n \geq \dfrac{WS - F}{T - S}$

46. One simply computes

$$\frac{WS - F}{T - S}$$

and takes the smallest integer, n^*, greater than or equal to this number. Adding one gives the index of the first "no-wait" customer.

47. $(W + n^*) \cdot S$

48. Yes. The process should look similar to the following:

$$(W + n) \cdot S \leq nT + F$$
$$WS - F \leq n(T - S)$$
$$\frac{WS - F}{T - S} \leq n.$$

In going from line two to line three, one is dividing both sides of an inequality by $T - S$. The sense of the inequality remains unchanged only when $T - S > 0$ or $T > S$.

49. If $F \geq WS$, then all of the customers waiting when the facility opens will be serviced before the arrival of any new customers. Thus the first new customer would be the first "no-wait" customer.

50. a. 0%
    b. 100%
    c. 2.5 minutes

51. a. 25%
    b. 0
    c. 25%
    d. 2.5 minutes

52. a. 1 customer
    b. 2 customers
    c. 3 customers
    d. 2 customers

53. a. 25%
    b. 50%
    c. 25%

54.

	Service Time	Customer
a.	3 min	1
b.	3 min	2
c.	1 min	3
d.	4 min	4
e.	3 min	5

55.

Arrivals

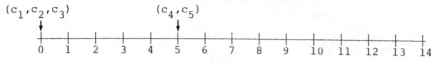

Start of Servicing

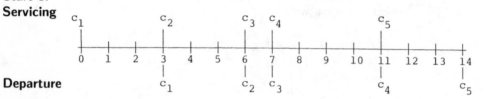

Departure

56. 3, 6

57. 1

58. In this sample, it was certain that the number of arrivals was 0 to 7.

59. 2.96

60. 10 or 20, 140 or 170 or 180 or 190 or 200 or 210.

61. 1

62. 62.6 seconds.

63. 4 customers

64. 74, 28

65. 4 arrivals, 1 arrival

66. 93

67. From page 25, values of 93–99 are assigned to arrival of 7 customers/minute.

68.

Random Number	Arrivals/Minute
46	3
31	2
85	5
33	2

69.

Customer	Random Number	Service Time
4	63	70
5	40	30
6	99	220
7	74	90

70.

Customer	Random Number	Service Time
88	80	110
89	72	90
90	42	40
91	60	60

71.

Customer	Arrival Time	Waiting Time	Start of Service	Service Time	Finish of Service	Clerk
11	180	260	440	30	470	*B*
15	240	300	540	10	550	*B*

72. 330 seconds

73.
a.	60 sec		**d.**	322 sec
b.	252 sec		**e.**	544 sec
c.	294 sec		**f.**	1062 sec

74.

Customer	Arrival Time	Waiting Time	Start of Service	Service Time	Finish of Service	Clerk
1	0	0	0	110	110	*A*
2	0	0	0	70	70	*B*
3	0	0	0	220	220	*C*
4	0	70	70	70	140	*B*
5	60	50	110	30	140	*A*
6	60	80	140	220	360	*A*
7	60	80	140	90	230	*B*
8	60	160	220	50	270	*C*
9	120	110	230	40	270	*B*
10	180	90	270	10	280	*C*
11	180	90	270	30	300	*B*
12	180	100	280	40	320	*C*
13	180	120	300	70	370	*B*
14	180	140	320	60	380	*C*
15	240	120	360	10	370	*A*

75.
   a. 24 sec
   b. 104 sec
   c. 114 sec

76.

Customer	Arrival Time	Waiting Time	Start of Service	Service Time	Finish of Service	Clerk
1	0	0	0	110	110	*A*
2	0	0	0	70	70	*B*
3	0	0	0	220	220	*C*
4	0	0	0	70	70	*D*
5	60	10	70	30	100	*B*
6	60	10	70	220	290	*D*
7	60	40	100	90	190	*B*
8	60	50	110	50	160	*A*
9	120	40	160	40	200	*A*
10	180	10	190	10	200	*B*
11	180	20	200	30	230	*A*
12	180	20	200	40	240	*B*
13	180	40	220	70	290	*C*
14	180	50	230	60	290	*A*
15	240	0	240	10	250	*B*

77.
   a. 2 sec
   b. 30 sec
   c. 26 sec